20TH CENTURY SCIENCE AND TECHNOLOGY

1970-90

COMPUTERS AND CHIPS

Please visit our web site at: www.garethstevens.com
For a free color catalog describing Gareth Stevens' list of high-quality books and
multimedia programs, call 1-800-542-2595 (USA) or 1-800-461-9120 (Canada).
Gareth Stevens Publishing's Fax: (414) 332-3567.

Library of Congress Cataloging-in-Publication Data

Parker, Steve.
    1970-90: computers and chips / by Steve Parker.
        p. cm. — (20th century science and technology)
    Includes bibliographical references and index.
    ISBN 0-8368-2946-8 (lib. bdg.)
    1. Technological innovations—History—20th century—Juvenile literature.
    2. Inventions—History—20th century—Juvenile literature. [1. Technological
    innovations. 2. Inventions—History—20th century. 3. Science—History—
    20th century. 4. Technology—History—20th century.] I. Title.
    T173.8.P36 2001
    509.047—dc21   2001020784

This North American edition first published in 2001 by
Gareth Stevens Publishing
A World Almanac Education Group Company
330 West Olive Street, Suite 100
Milwaukee, WI 53212 USA

Original edition © 2000 by David West Children's Books. First published in Great Britain
in 2000 by Heinemann Library, Halley Court, Jordan Hill, Oxford OX2 8EJ, a division of
Reed Educational and Professional Publishing Limited. This U.S. edition © 2001 by
Gareth Stevens, Inc. Additional end matter © 2001 by Gareth Stevens, Inc.

Designers: Jenny Skelly and Aarti Parmar
Editor: James Pickering
Picture Research: Brooks Krikler Research

Gareth Stevens Editor: Valerie J. Weber

Photo Credits:
Abbreviations: (t) top, (m) middle, (b) bottom, (l) left, (r) right

Apple Macintosh: cover(b), pages 4–5(b), 22–23(t).
British Aerospace Airbus: page 18(m).
British Airways: page 24(t).
Corbis Images: pages 4(t), 11(tl), 11(tr), 14 both, 15(t), 22(bl), 27(bl), 27(br).
European Space Agency: page 11(br).
Frank Spooner Pictures: pages 6(t), 16(tl), 16-17, 21(b), 23(b), 25(b), 26(r), 27(t).
JVC: page 5(t).
NASA: pages 5(m), 5(b), 8 all, 9(t), 10 all, 11(ml), 11(mr), 11(bl), 12 all, 13(b), 17(t), 23(m).
Oxford Scientific Films: page 6(b).
Peugeot: page 24(b).
Rex Features: pages 7(t), 16(b), 17(b), 19 both, 20-21, 24-25(t), 26(l), 28 both, 29(tr), 29(br).
Sandia: page 21(t).
Sinclair: page 25(t).
Solution Pictures: cover (m), pages 18-19(t), 29(tl).

Printed in the United States of America

1 2 3 4 5 6 7 8 9 05 04 03 02 01

20TH CENTURY SCIENCE AND TECHNOLOGY

1970-90

COMPUTERS AND CHIPS

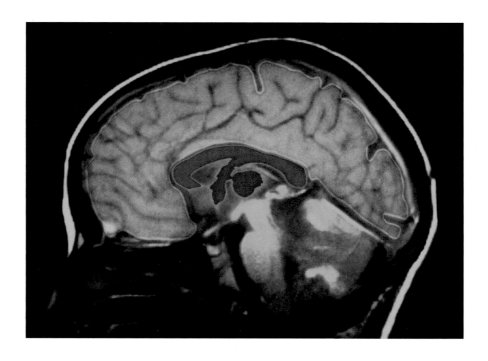

Steve Parker

Gareth Stevens Publishing
A WORLD ALMANAC EDUCATION GROUP COMPANY

CONTENTS

TWO SIDES OF SCIENCE 5

NEW VIEWS 6

OBSERVING EARTH 8

EXPLORING SPACE 10

SPACE STATIONS 12

SPACE SHUTTLES 14

SCIENCE ON TRIAL 16

ON THE MOVE 18

ALTERNATIVE ENERGY 20

COMPUTERS 22

TRAVEL-TECH 24

MEDICAL SCIENCE 26

GADGETS 28

TIME LINE 30

GLOSSARY 31

BOOKS AND WEB SITES 31

INDEX 32

The space shuttle, a new kind of reusable spacecraft, first blasted off in 1981. In 1986, however, a disaster set the shuttle program back many years.

Apple computers, such as the early Macintosh, pioneered easy-to-use on-screen lists, or menus, of various options.

TWO SIDES OF SCIENCE

Science and technology made great strides during the 1960s, climaxing with the 1969 Moon landing. In the 1970s, the pace of advances in these fields sped up, especially in electronics and computers. The negative side to all this progress, however, was increasingly revealed during the 1970s and 1980s. New scientific developments and technical innovations had brought new problems.

In 1975, astronauts from rival superpowers, the Soviet Union and the United States, shook hands in space.

The 1980s saw the spread of handheld video cameras, allowing people to make home videos.

Pollution became world news, with oil slicks on coastlines, smog over cities, leaks from chemical factories and nuclear power stations, acid rain, the hole in the ozone layer, and evidence of global warming. These environmental problems were not a result of science itself but of the way it was being applied. Seeking to control the way science and technology were being used, people formed organizations to increase safety, protect the environment, and conserve natural resources.

The Soviet Mir space station was launched in 1986 and abandoned in 1999.

5

NEW VIEWS

Scientists developed exciting theories and explored many remote places during the 1970s, from the farthest reaches of outer space to the bottom of the sea here on Earth.

INNER SPACE

The three-person deep-sea craft *Alvin*, launched in 1964, made some exciting discoveries. In 1977, it found unexpected life on the deep-sea bed in the eastern Pacific Ocean. Giant worms as thick as human arms, shellfish the size of dinner plates, blind crabs, and creeping fish clustered around cracks in the ocean floor where hot, dark, mineral-rich water spurted out from far below.

In 1986, the submersible Alvin *explored the wreck of the ocean liner* Titanic, *which sank in the northwestern Atlantic in 1912.*

NEW LIFE

How do these eerie deep-sea creatures survive in the cold, black ocean? Microbes, especially bacteria, in these animals absorb the energy-rich minerals from the water. They use the minerals for growth and, in turn, supply energy and nutrients for the animals to use. On land and in shallow seas, animals eat plants or other plant-eating animals — and plants need sunlight to grow. The animals living in the deep-sea cracks don't depend on the Sun. This discovery raised the possibility of life on other worlds, life powered by the energy in minerals rather than light from stars.

Huge worms and ghostly crabs crowd around a "black smoker," a dark plume of mineral-rich water flowing through a crack or vent in the deep-sea bed.

BLACK HOLES ARE NOT QUITE BLACK

A black hole is a place where atoms and other forms of matter are squeezed into a tiny region, packed together so tightly that their density is infinite. The black hole's gravitational pull is so strong that nothing, not even light, can escape it, which is why the hole is black. Matter, such as dust or even a planet that comes near the black hole, is sucked in and seems to disappear. The basic idea of black holes goes back to the theories made by a French mathematician, Pierre Simon de Laplace, in 1798. In the early 1900s, Albert Einstein helped explain them with his theory of relativity. In the 1970s, scientist Stephen Hawking made more advances in our understanding of black holes.

Stephen Hawking (b. 1942)

a black hole, a region where space itself bends or curves like a bottomless pit

dust and other matter streaming from a star

star

matter sucked into a black hole

Hawking found that under certain conditions, a black hole could emit tiny quantities of heat energy. The black hole then becomes a faintly glowing red hole. Small black holes may occur almost anywhere when solid matter the size of a mountain is squeezed into a space smaller than one atom. Galaxies, massive swirling groups of billions of stars, may have giant black holes at their centers.

IRAS, the Infrared Astronomical Satellite, detected sources of infrared radiation from deep space. In 1983, it found a new star type, the cool brown dwarf.

OBSERVING EARTH

While landing on the Moon was the focus of the 1960s, the 1970s became the decade of satellite launches. Dozens of research, weather, and survey satellites were put into orbit to look down on Earth and out into space.

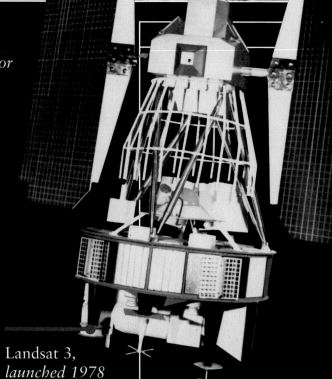

Launched in June 1978, Seasat (Specialized Experimental Applications Satellite) was 39 feet (12 meters) long.

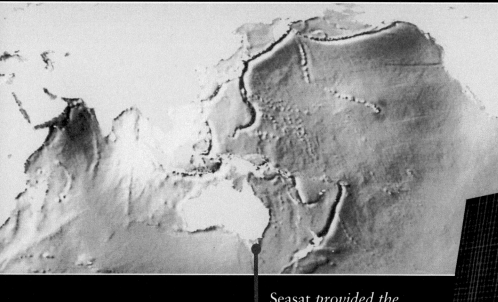

Seasat provided the information for this map of the ocean floor showing deep-sea trenches.

LAND AND SEA

Starting in 1972, the United States launched a series of *Landsat* satellites into specialized orbits. The satellites carried cameras for taking ordinary photographs as well as for capturing images of heat and other types of energy. The detailed photographs are used by farmers studying crop fields, environmental experts looking at habitat destruction, and geologists searching for oil, coal, and other resources. For four months, another satellite, *Seasat*, provided information about the oceans.

Landsat 3, *launched 1978*

WHAT'S THE WEATHER?

Meteorological satellites such as the *Nimbus* and *Meteosat* series carried cameras to photograph clouds, storms, and other weather features. Each *Nimbus* was about 10 feet (3 m) long and weighed up to 2,210 pounds (1 metric ton). It took about 1,500 photos every day. The satellites' sensors detected temperature, humidity, wind speed, and similar conditions. The vast amount of information they gathered helped forecasters predict the weather and also showed how the global climate may be changing because of the greenhouse effect. *Nimbus 3* was launched in 1970. The last of the series, *Nimbus 7*, went into orbit in 1978. It was the first satellite designed to detect polluting chemicals in the atmosphere.

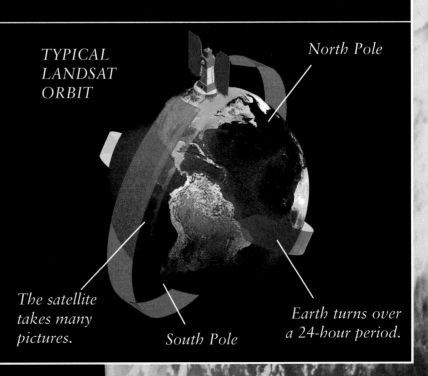

drum surface covered with solar cells that change sunlight into electricity

Meteosat 1 *was launched in 1974 and* Meteosat 2 *in 1975. Each drum-shaped satellite measured 6 feet (2 m) by 11 feet (3.5 m) and weighed 541 pounds (245 kg). They were put into geosynchronous orbit, completing one orbit every twenty-four hours.*

9

SPECIAL ORBITS

Satellites have different orbits around Earth depending on their job. *Landsats* have low polar orbits, traveling over the North and South Poles and taking strips of images as Earth turns. At their closest, the satellites are less than 560 miles (900 kilometers) above the surface, so their cameras can provide amazing details. The military uses *Landsats* and similar satellites to check on the movements of warships, fighter planes, tanks, and even soldiers.

TYPICAL LANDSAT ORBIT

North Pole

The satellite takes many pictures.

South Pole

Earth turns over a 24-hour period.

EXPLORING SPACE

A series of unpiloted, deep-space probes launched in the 1970s and 1980s vastly improved our knowledge of the Sun and planets in the Solar System.

Mariner 10 took close-up photos of Mercury and Venus, and discovered Mercury's natural magnetic field.

MERCURY

WE ARE HERE

The *Voyager* probes carried plaques showing the likeness of a man and woman and a map of Earth's position in space. One day, an alien intelligence may find the probe and perhaps visit us.

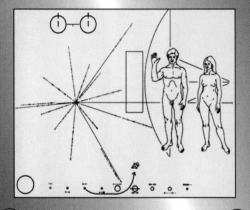

Voyager's *gold-plated plaque*

INNER PLANETS

Mariner probes were designed to study the planets relatively near Earth and close to the Sun — Mercury, Venus, and Mars. In 1971, *Mariner 9* traveled to Mars and took spectacular photos of its huge mountains and vast valleys. *Mariner 10* blasted into space in 1973 and passed within 3,586 miles (5,770 km) of Venus, photographing its dense clouds. Using the gravitational pull of the planet like a slingshot, the probe shot onward to circle Mercury three times.

VENUS

10

LIFE ON MARS?

In 1975, two U.S. spacecrafts, *Vikings 1* and *2*, were sent to explore Mars. They arrived in 1976. Each craft separated into an orbiter and a lander. The orbiters circled the planet, taking photographs and sensor readings, while the landers parachuted to a soft touchdown on the surface. The landers also took hundreds of photographs, measured gases in the atmosphere, and scooped up samples of the reddish soil to test for signs of life — but found none.

1. Orbiter and lander separate.

5. Orbiter receives signals from lander and sends them to Earth.

2. Thrusters position lander for entry into atmosphere.

3. Lander's parachute opens at 19,680 ft. (6,000 m).

4. Soft landing as sensors on "feet" shut down engines.

Viking *lander*

JUPITER

SATURN

NEPTUNE

OUTER PLANETS

Launched in 1977, U.S. probes *Voyager 1* and *2*, journeyed to the planets farther out in the Solar System. *Voyager 1* passed within 167,806 miles (270,000 km) of Jupiter in 1979 and traveled within 74,580 miles (120,000 km) of Saturn in 1980. *Voyager 2* also flew past Jupiter and Saturn, coming within 49,720 miles (80,000 km) of Uranus in 1986. Three years later, it soared over Neptune, passing only 3,108 miles (5,000 km) away. Both *Voyager* probes have now left the Solar System and are speeding through interstellar space.

Each Voyager *probe carried an antenna 13 feet (4 m) in diameter to send and receive Earth's radio signals.*

11

MARS

The Giotto *probe, launched in 1985 by a European Ariane rocket, passed Halley's comet in 1986. It discovered that the comet's ice and dust core is only 6 miles (10 km) across.*

SPACE STATIONS

After making short trips into space, the astronauts' next great challenge was to live there for weeks, months, or maybe even years. Could the human body cope with conditions such as weightlessness, lack of exercise, and the feeling of isolation?

Skylab *needed a replacement gold sunshade after an accident at launch.*

U.S. and Soviet astronauts train in a mock-up of the Soviets' Soyuz *craft in early 1975.*

SALYUTS

As in the 1960s, the two superpowers — the United States and the Soviet Union (now mainly Russia) — raced each other to develop space stations. The Soviets won: *Salyut 1* went into orbit on April 19, 1971, to mark the tenth anniversary of the first piloted space flight by Yuri Gagarin. Over the next eleven years, six more *Salyuts* were launched. The last, *Salyut 7*, went into orbit in 1982. It was boosted into a higher orbit in 1986 so it would not reenter Earth's atmosphere and crash to Earth like earlier *Salyuts*. This move did not work; parts of *Salyut 7* fell on Argentina.

SKYLAB

The U.S. *Skylab* space station blasted off in 1973, housed in a modified *Saturn V* rocket casing. Soon after launch, one of its two solar panels was torn off, and the other panel and a shield to protect against the Sun's rays were damaged. Four crews visited the station in *Apollo* spacecraft over nine months, carrying out repairs and many scientific experiments.

A symbol of superpower cooperation, astronauts meet as Apollo 18 *and* Soyuz 19 *dock in orbit in July 1975.*

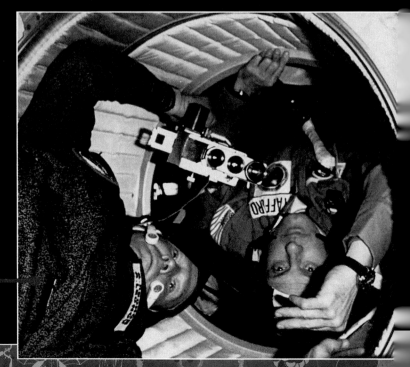

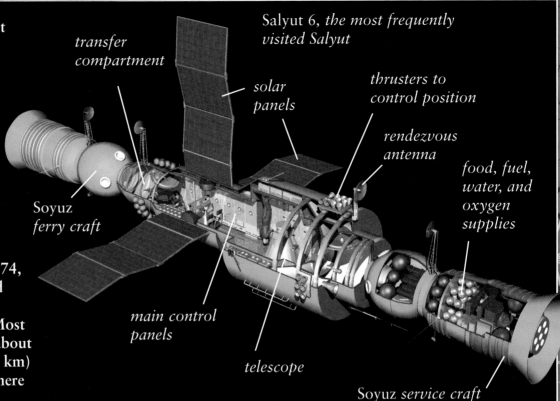

Each *Salyut* was about 44 feet (13.5 m) long and weighed nearly 20 tons (18 m tons). It was designed for two people with occasional extra visitors. The first crew members entered *Salyut 1* in June 1971 and stayed for twenty-two days. They died, however, when a pressure valve failed as they returned to their ferry craft for the journey back to Earth. *Salyut 3*, sent up in 1974, and *Salyut 5*, in 1976, carried out military experiments and photography for espionage. Most Salyuts orbited at heights of about 155 to 217 miles (250 to 350 km) and reentered Earth's atmosphere as "shooting stars" in a year.

transfer compartment

Salyut 6, *the most frequently visited Salyut*

solar panels

thrusters to control position

rendezvous antenna

food, fuel, water, and oxygen supplies

Soyuz ferry craft

main control panels

telescope

Soyuz *service craft*

MIR

The Soviets launched their space station *Mir* (Russian for "peace") into orbit in 1986. Its basic structure was larger than *Salyut*, and *Mir* also had better facilities — more windows, two private compartments, and extra hatches for linking to other craft. The unpiloted *Progress* craft brought basic supplies to *Mir, Kvant* carried scientific equipment and experiments, and the *Soyuz TM* ferried the cosmonauts.

In 1988, Soviet cosmonauts Musa Manarov and Vladimir Titov became the first people to spend a year in space on board Mir. In this view, the station core is vertical with various ferry craft, including Kvant and Soyuz, attached in the middle.

SPACE SHUTTLES

With the U.S. space shuttles, a new era in space began in 1981. The shuttles did not burn up on reentry or drift away from Earth; they flew back down to be used again.

REUSABLE SPACECRAFT

A space shuttle has four main parts. One is the spacecraft itself, the orbiter. At launch, a giant fuel tank 154 ft. (47 m) tall supplies the orbiter's three rocket engines with liquid oxygen and fuel. Two solid fuel rocket boosters cling to either side of the tank. The boosters and tank fall away as the orbiter approaches its maximum height. Five orbiters were built — *Enterprise, Columbia, Challenger, Atlantis* and *Discovery. Columbia* was first into space on April 12, 1981.

When the shuttle is in orbit, the doors on the orbiter's payload bay open so satellites or other objects can be released.

The shuttle blasts off using two solid rocket boosters (SRBs) and an external fuel tank. Non-reusable, the tank burns up in the atmosphere, while the SRBs parachute back to Earth.

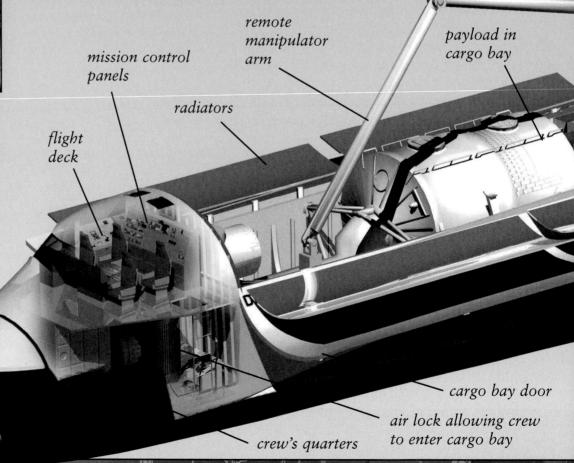

mission control panels

remote manipulator arm

payload in cargo bay

radiators

flight deck

forward thrusters for maneuvering in space

forward landing gear

cargo bay door

air lock allowing crew to enter cargo bay

crew's quarters

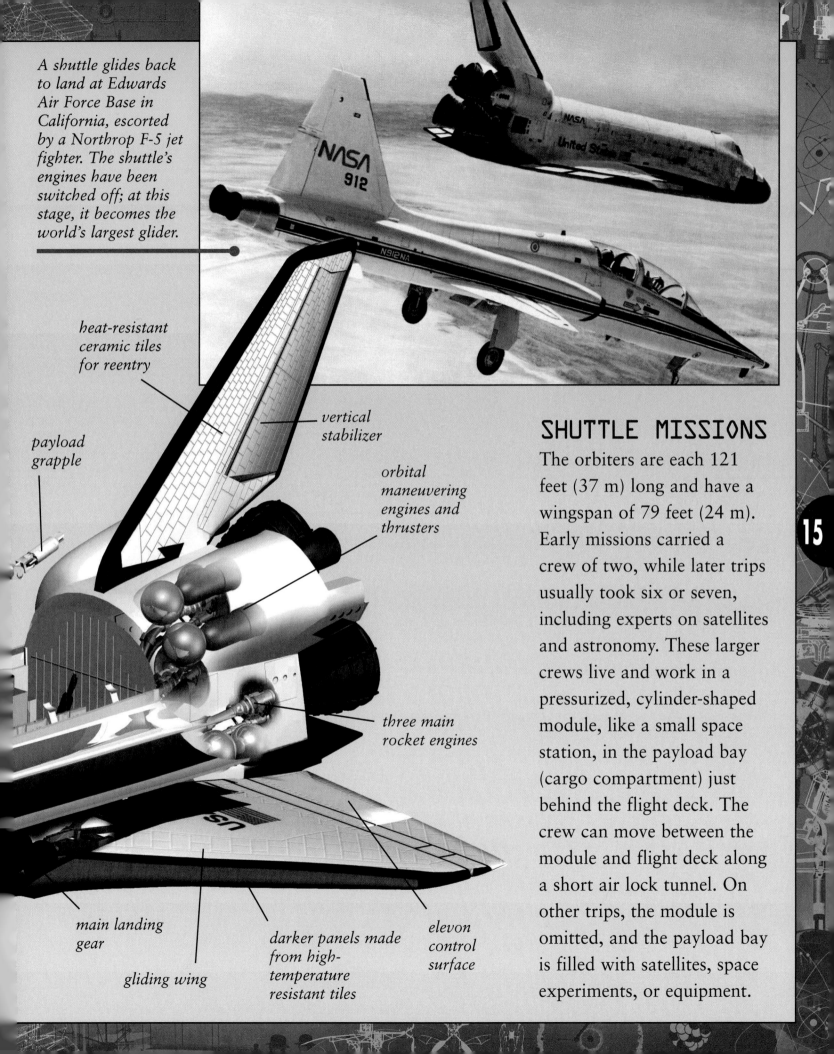

A shuttle glides back to land at Edwards Air Force Base in California, escorted by a Northrop F-5 jet fighter. The shuttle's engines have been switched off; at this stage, it becomes the world's largest glider.

heat-resistant ceramic tiles for reentry

payload grapple

vertical stabilizer

orbital maneuvering engines and thrusters

three main rocket engines

main landing gear

gliding wing

darker panels made from high-temperature resistant tiles

elevon control surface

SHUTTLE MISSIONS

The orbiters are each 121 feet (37 m) long and have a wingspan of 79 feet (24 m). Early missions carried a crew of two, while later trips usually took six or seven, including experts on satellites and astronomy. These larger crews live and work in a pressurized, cylinder-shaped module, like a small space station, in the payload bay (cargo compartment) just behind the flight deck. The crew can move between the module and flight deck along a short air lock tunnel. On other trips, the module is omitted, and the payload bay is filled with satellites, space experiments, or equipment.

SCIENCE ON TRIAL

In 1976, a cloud of poisonous gas accidentally escaped from a chemical factory in Seveso, Italy. Many people blamed science for this disaster and later catastrophes.

An aerial view of the terrible destruction of the nuclear reactor at Chernobyl.

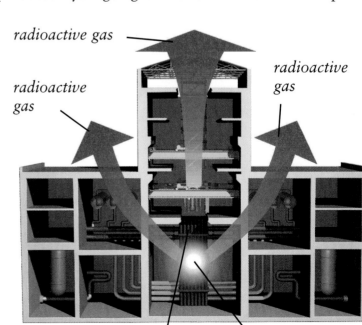

SEVESO

The Seveso cloud contained dioxins, dangerous chemicals that are by-products of making herbicides. Dioxins can pollute the soil and harm living things. Farm animals, dogs, and other pets died at Seveso, but no people. Still, the accident made many people aware of the problems that science-based processes might cause.

Many victims in Bhopal suffered blindness.

16

BHOPAL

In 1984, another leak at a chemical factory released a cloud of gas over the city of Bhopal, India, injuring hundreds of people. The accident raised questions about whether it was wise to build dangerous factories in cities.

Workers wearing protective suits and masks prepare to clean up the Seveso site in Italy.

NUCLEAR DISASTER

The Chernobyl accident began with a cooling-pipe leak near the nuclear core's base. Water poured out, reacting with the graphite that helps control nuclear fission. The reaction produced hydrogen gas that collected and then exploded.

radioactive gas

radioactive gas

radioactive gas

nuclear reactor

explosion

CHERNOBYL

In 1986, a small leak in a water pipe set off a chain of events that caused a terrible tragedy — an explosion in the Number 4 nuclear reactor at the

Chernobyl power station near Kiev, Ukraine. The explosion and its aftermath killed thirty-one people. It also allowed radioactive gas to escape from the huge building, drift away, and pollute an enormous area, contaminating soil, farm crops, and animals for several years.

MORE TRAGEDIES

That same year, the space shuttle *Challenger* blew up shortly after liftoff, killing all seven on board. In 1989, the supertanker *Exxon Valdez*

New fertilizers and machines increased crop yields. As grain piled up in rich lands, however, millions in poor regions starved.

spilled huge amounts of oil into Alaskan waters. The oil slick killed millions of birds, seals, whales, and other sea life. In addition, during the 1980s, scientists discovered the ozone layer was thinning. This blanket of ozone gas (a form of oxygen) in the upper atmosphere helps absorb some of the Sun's harmful ultraviolet rays. With less ozone, more rays could reach Earth.

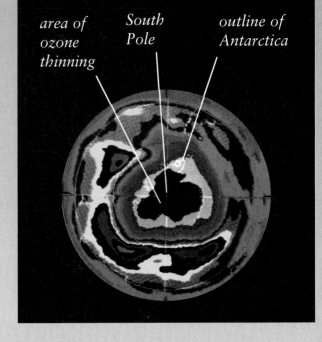

ON THE MOVE

The shrinking size of electronics meant many processes could be automated, even flying a jetliner.

RIVALS

In 1970, several European nations joined together to form Airbus Industries. Their plan was to build large jet passenger aircraft that could rival those of huge U.S. plane manufacturers such as Boeing and Lockheed. Their first plane, the Airbus *A300*, flew in 1972.

The use of industrial robots, ideal in the difficult conditions of a car factory, became wide-spread in the mid 1970s.

18

EUROPLANES

Airbus planes are a cooperative effort between many European countries. Different parts, such as the wings, fuselage, tail, and engines, are built in different places, then shipped to Toulouse, France for final assembly.

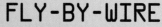

☐	*Spain*
■	*Belgium*
☐	*France*
☐	*Germany*
☐	*England*

Plans for the Airbus A340 were approved in 1987; it began serving passengers in 1993.

FLY-BY-WIRE

Airbus pioneered the "fly-by-wire" system. Cables or pipes once linked the main controls on the flight deck to the aircraft's parts, such as the rudder on the tail fin. With fly-by-wire, the plane's computer senses the controls' movements, converts them into electrical signals, and sends the signals along wires to motors that move the parts. It also warns the crew of problems.

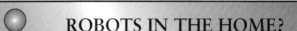

Australia's Sydney Opera House is designed to look like a ship's sails billowing in the wind or shells piled up on the seashore.

NEW MATERIALS

Researchers continued to develop new building materials, including different types of steel, concrete, and carbon fiber composites. The strength and adaptable nature of these materials allowed designers to create spectacular shapes and effects. With a series of vast, curved roofs made of reinforced concrete covered with gleaming white ceramic tiles, the Sydney Opera House, opened in 1973, stands on a narrow strip of land jutting into Sydney Harbor. In 1981, the Humber Bridge in England became the longest single bridge span, at 4,625 feet (1,410 m).

ROBOTS IN THE HOME?

During the 1970s, computer-controlled robotic machines appeared on many production lines. After being programmed to mimic human movement, they carried out the same motions precisely, every time, without becoming tired or distracted. More intelligent robots to help with household chores, however, were still a dream.

A robot waiter clears away dirty dishes — funny but not very practical!

19

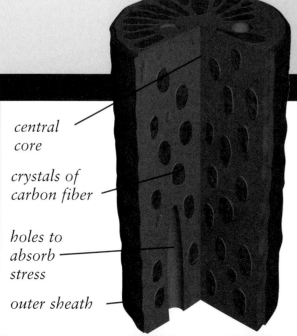

central core

crystals of carbon fiber

holes to absorb stress

outer sheath

FIBERS + RESINS = STRENGTH

Carbon fiber composites are made from fibers of a carbon-based substance, such as artificial rayon or acrylic, and a rubbery, flexible, resin-based material. The resulting composite is five times stronger than steel, but it bends to absorb stresses rather than cracking.

In carbon fiber laminate, layers of fibers are at different angles.

ALTERNATIVE ENERGY

In 1973, oil-producing nations raised oil prices and restricted supplies, shocking oil-dependent countries into reevaluating the energy they used.

WAVE POWER

Could wave energy be harnessed for electricity? Many designs were tried, including Salter's ducks, which rock up and down to rotate a central axle. No design, however, has proved practical. Huge storm waves usually damage the models.

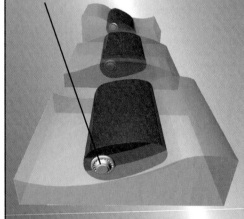

central axle

Salter's duck design (1974)

THE OIL CRISIS

The sudden leap in the price of oil made industrial regions realize how much they depended on this valuable resource. Oil is used in power stations and processed to make fuels, lubricants, plastics, chemicals, and hundreds of other products. To lessen this dependence, reduce the pollution oil causes, and make supplies last longer, the search began for other sources of energy.

By 1980, the Hoover Dam's seventeen generators were providing 2,000 megawatts of electricity.

HYDROELECTRICITY

To produce hydroelectricity, flowing water spins the angled blades of a turbine that is linked to an electricity generator. To increase the pressure and flow of water and make its availability more reliable year-round, these power stations are built in dams across large rivers. Once built, a hydroelectric power station needs no fuel, produces no air pollution, and has low maintenance costs.

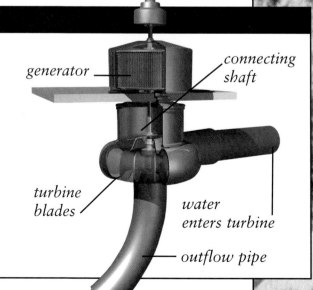

generator

connecting shaft

turbine blades

water enters turbine

outflow pipe

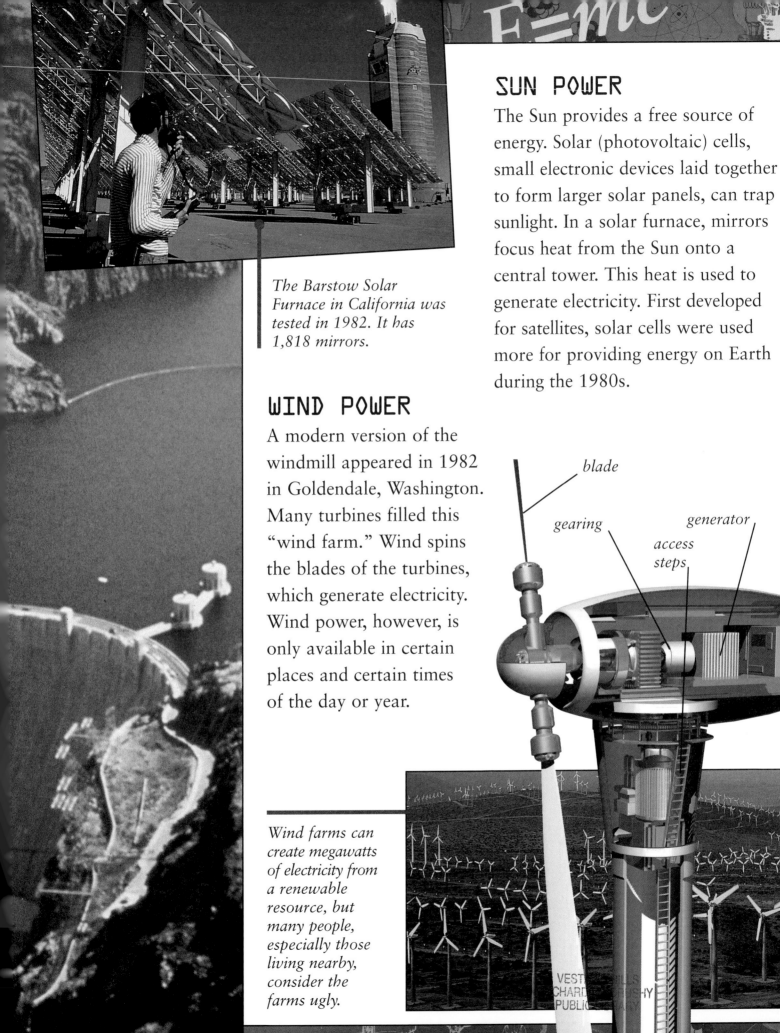

SUN POWER

The Sun provides a free source of energy. Solar (photovoltaic) cells, small electronic devices laid together to form larger solar panels, can trap sunlight. In a solar furnace, mirrors focus heat from the Sun onto a central tower. This heat is used to generate electricity. First developed for satellites, solar cells were used more for providing energy on Earth during the 1980s.

The Barstow Solar Furnace in California was tested in 1982. It has 1,818 mirrors.

WIND POWER

A modern version of the windmill appeared in 1982 in Goldendale, Washington. Many turbines filled this "wind farm." Wind spins the blades of the turbines, which generate electricity. Wind power, however, is only available in certain places and certain times of the day or year.

blade

gearing

generator

access steps

Wind farms can create megawatts of electricity from a renewable resource, but many people, especially those living nearby, consider the farms ugly.

COMPUTERS

In today's industrial world, a computer sits in almost every office and home. In 1970, however, computers were big and costly and were used only by large businesses, universities, and governments.

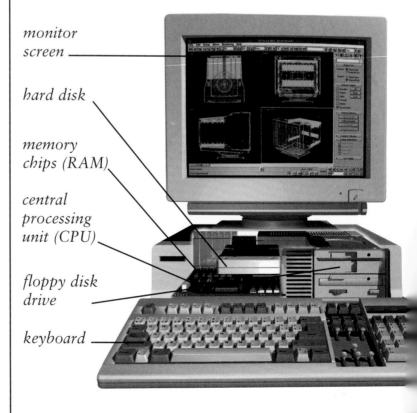

The Apple Macintosh range of small computers began in the 1980s.

COMPUTER PROGRESS

The era of the personal computer, small and cheap enough to buy for home use, began in 1975 with the Altair 8800. This computer was advanced for its time but not especially successful. The Apple II followed in 1977 and was much more popular. In 1981, IBM introduced its first personal computer (PC), with floppy disks and many other features still used today.

In 1985, a newly developed type of fiber optic increased the carrying capacity of one hair-thin fiber to 300,000 telephone calls.

THE COMPUTER SET-UP

The basic personal computer took shape during the late 1970s and early 1980s. The 1983 Apple Lisa was the first to have a mouse controlling a pointer on the screen with a button to click on different choices or options. It also had the pull-down lists, or menus, familiar today. Though PCs have not changed much in size or shape since then, they double in power, memory, and processing speed every eighteen to twenty-four months.

monitor screen

hard disk

memory chips (RAM)

central processing unit (CPU)

floppy disk drive

keyboard

GLOBAL COMMUNICATION

In December 1980, the first of a new generation of satellites was blasted into orbit. *Intelsat V* was a communications satellite (comsat) of the International Telecommunications Satellite Organization.

More than one hundred nations became involved in the organization, and fourteen more *Intelsat Vs* were launched over the following years. Their main job was to detect radio signals and microwaves — carrying telephone calls, TV programs, and computer data — sent from one part of Earth, boost them to higher strengths, and send them down to another part of Earth thousands of miles (km) away. The comsats allowed TV programs to be beamed live to almost every region of the world.

Each Intelsat V *could carry 12,000 phone calls and two TV channels at the same time.* Intelsats *relayed worldwide TV events such as the 1985 Live Aid concert for famine relief.*

The mouse transfers hand movements similar to those made in drawing onto the screen. As the mouse moves, the rubber ball inside rolls and spins sensors that track the mouse's motion.

roller sensors that pick up movement and feed it to the computer

click switch

rolling ball

23

TRAVEL-TECH

The oil crisis of the 1970s put a global brake on developing bigger, faster types of travel. The focus of research changed to reducing noise, waste, and pollution.

The supersonic Concorde, which could cruise at twice the speed of sound, suffered from high fuel prices. People preferred to pay more in time but less in cash.

24

SLOWDOWN

Many cities suffered from massive traffic jams and blankets of smog caused by vehicles belching fumes. They started to plan rapid mass transit systems to carry many people quickly, quietly, and safely, with minimal waste and pollution.

A monorail wends through Sydney, Australia. Overhead electric monorails pass above existing streets and are energy efficient and quiet.

An electric car is nearly three times more energy efficient than a gas-driven car. This 1984 electric car's many batteries, however, made it too heavy and slow.

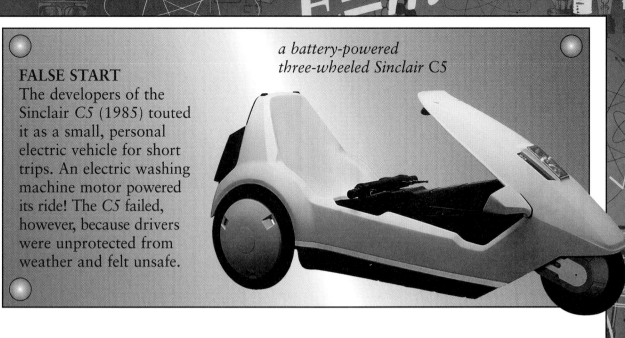

FALSE START
The developers of the Sinclair C5 (1985) touted it as a small, personal electric vehicle for short trips. An electric washing machine motor powered its ride! The C5 failed, however, because drivers were unprotected from weather and felt unsafe.

AN ATTRACTIVE SOLUTION

The maglev train, developed in the 1970s in Japan, Germany, and Britain, was a unique train. The *maglev,* which means "magnetic levitation," uses magnetism. A magnet has two regions of its strongest magnetic force, its north and south poles. Like poles repel each other; opposite poles attract. Most maglev trains have magnets with one pole facing the track; the same pole on the track's magnet faces the train. The two poles repel, and the train floats above the track, held up by magnetic force. The main problem with maglevs is the cost of a track full of magnets.

HOW A MAGLEV WORKS

One set of magnets makes the train rise, reducing noise and the energy wasted through friction in an ordinary train system. The other set switches on and off rapidly to attract the train from the front and repel it from behind, moving the train forward.

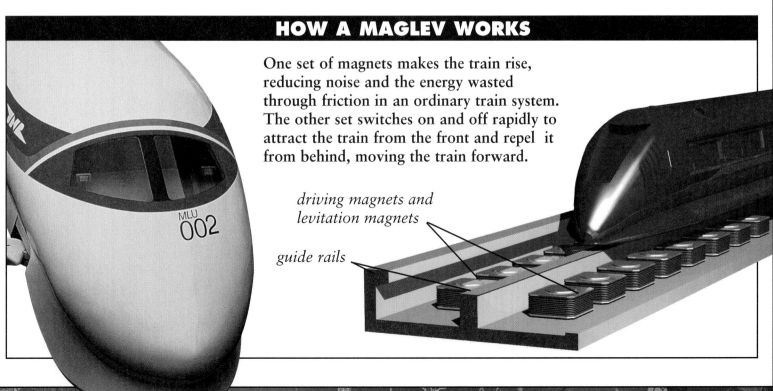

driving magnets and levitation magnets

guide rails

MLU 002

MEDICAL SCIENCE

The electronics revolution greatly affected medical technology as a new generation of equipment began to appear in hospitals and medical clinics.

SEEING INTO THE BODY

One of the major advances was in body imaging — seeing inside the body without cutting it open. Developed in 1972, the computerized tomography (CT) scanner uses very weak X rays to take pictures of thin "slices" of the body and combine the pictures into a three-dimensional image. Completed in 1973, the magnetic resonance imaging (MRI) machine places the body in a very strong magnetic field and fires tiny radio pulses through it. Both the CT and the MRI relied on the processing power of the new computers.

Barney Clarke received the first permanent artificial heart in 1982. He survived 112 days after the surgery.

SEEING THE INVISIBLE

Another leap forward was the advanced electron microscope. Instead of light rays, this microscope uses beams of electrons to magnify objects a million times or more. Scientists used it to identify the Human Immunodeficiency Virus (HIV), which causes Acquired Immune Deficiency Syndrome (AIDS).

HIV, the virus causing AIDS, was identified in 1984.

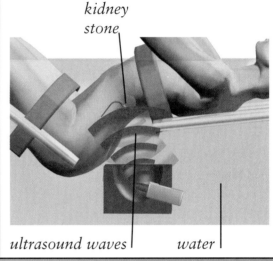

Today, IVF is a routine procedure, but Louise's arrival in 1978 was a major event and turned her into a celebrity.

TEST TUBE BABIES

The first "test tube" baby, Louise, was born to mother Lesley Brown in 1978 with the use of IVF, *in vitro* ("in glass") fertilization. In this process, a fiberoptic instrument called a laparoscope harvests tiny eggs from the mother or a donor. The eggs are mixed with sperm from the father or another donor in a glass dish — not really a test tube. Sperm and egg join and begin to develop into an embryo, which is put into the mother's uterus to grow into a baby.

In 1980, a new type of medical machine began to shatter hard, stony lumps that sometimes form inside the body, usually inside the gall bladder or kidneys. The lithotripsy machine fires powerful high-pitched sound waves at the stone. They make the stone vibrate and break into many tiny pieces that leave the body naturally.

kidney stone

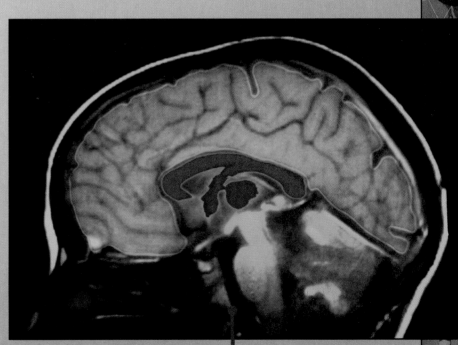

ultrasound waves   *water*

27

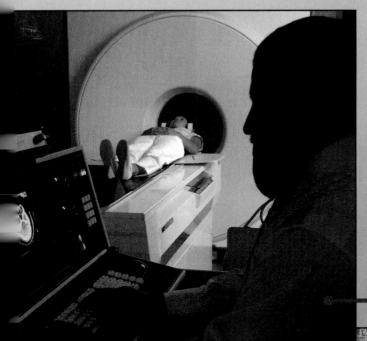

An MRI gives doctors clear and detailed pictures without X rays.

Computers often color MRI and CT scans to show details more clearly.

GADGETS

A gadget revolution began in the 1970s. It was based on electronic integrated circuits (ICs), known as silicon chips, microchips, or simply chips.

VERSATILE CHIPS

A microchip is a small sliver of silicon, just a few hundredths of an inch (millimeters) square, with many microscopic lines, patterns, and shapes etched on its surface. Silicon is a semiconductor: sometimes it carries or conducts electricity well, in other conditions, it does not. The patterns on the chip use this feature to work as resistors, transistors, and other electronic components. They make circuits that manipulate pulses of electricity at incredible speeds.

28

As the processing speed of microchips advanced, they could control increasingly more complex and fast-moving video games.

The Sony Walkman personal radio cassette player (1979) entertained joggers, travelers, and anyone on the go.

ON THE MOVE

The components on a microchip are already connected or integrated into circuits (thus, the term *IC*) instead of linked later by wires. Chips are tiny, light, tough, and use little electricity so they only need small batteries, making them ideal for portable electronic gadgets.

The Post-it® Note (1980) turned a glue too weak for ordinary use into a success.

WORK AND PLAY

Some small, lightweight gadgets, such as tape recorders and pocket calculators, were mainly for work. Others, such as personal music players and hand-held video games, were for entertainment. The liquid crystal display (LCD) played a vital part in the gadget revolution. It shows patterns of dark shapes on a clear background depending on the electronic signals it receives. Like the IC, it is tiny, tough, and uses little electricity.

Mobile phones had begun to appear by the end of the 1980s, although they were expensive and bulky.

THE LCD

An LCD is a sandwich of units including crystals and polarizing filters. The crystals can make a dark area by twisting light rays so they do not pass through the filter or reflect off the mirror at the base of the LCD.

The 1989 Nintendo Game Boy's LCD screen showed small, simple moving pictures.

No light reflected produces a dark area.

Light reflected produces a light area.

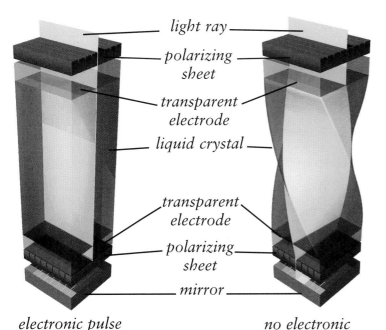

light ray

polarizing sheet

transparent electrode

liquid crystal

transparent electrode

polarizing sheet

mirror

electronic pulse received

no electronic pulse received

TIME LINE

	WORLD EVENTS	SCIENCE EVENTS	TECHNOLOGY	FAMOUS SCIENTISTS	INVENTIONS
1970	•U.S. troops sent into Cambodia	•Apollo 13 mission cut short, crew saved	•First video cassette recorders	•Stephen Cook shows many logic problems are one	•Removable floppy computer disc
1971	•Uganda: Idi Amin seizes power	•First microprocessors, tiny electronic "brains"	•USSR launches first space station, Salyut 1	•Niklaus Wirth's PASCAL computer language	•Food processor
1972	•Direct rule in Ulster •U.S. troops leave Vietnam	•U.S.: Massive new atom-smasher, Batavia	•First pocket calculator	•Murray Gell-Mann links quantum theory and quarks	•Home video game, the bat-and-ball "Odyssey"
1973	•Yom Kippur War •OPEC raises oil prices	•Pioneer 10 at Jupiter	•U.S. launches Skylab space station	•E. Tryon suggests Universe could start from nothing	•Push-in ring-tabs for drink cans
1974	•Turkey invades Cyprus and occupies one-third	•Signs of ozone damage become clearer	•First practical wave-powered generator	•Don Johanson and team find ancient "Lucy" fossils	•Bar code laser scanners used in retail stores
1975	•Cambodia overrun by Pol Pot's Khmer Rouge	•European Space Agency formed	•French company, BIC, invents disposable razors	•John Cornforth's Nobel Prize for work on enzymes	•First home computer available in kit form
1976	•South Africa: Soweto uprising	•Guidelines agreed on for genetic engineering	•The Concorde's first commercial flight	•Khorana and team make an artificial gene	•Ink-jet printer •Fiber-optic telecom cables
1977	•UN bans arms sales to South Africa	•Somalia: Last natural case of smallpox	•First human-powered flight, in Gossamer Condor	•Efron's "bootstrap" high-speed computer statistics	•Jobs and Wozniak's Apple II home computer
1978	•Egypt and Israel sign Camp David peace treaty	•Electron beams make microchips even smaller	•Apple disc drive for small computers	•Christy and Harrington find Pluto's moon, Charon	
1979	•Iran: Khomeini in power •USSR invades Afghanistan	•U.S.: Three Mile Island nuclear accident	•Computer spreadsheets •Vehicle exhaust CATs	•Jean Ichbia develops ADA computer language	•Walkman personal stereo •Videocassettes widespread
1980	•Start of Iran-Iraq War •Poland: Solidarity	•Scientists record Mt. St. Helens volcanic eruption	•Scanning tunneling microscope sees one atom	•Alan Guth suggests inflationary Universe idea	•Erno Rubik's cube sparks puzzle craze
1981	•U.S.: Ronald Reagan becomes President	•Stealth fighter flights	•First space shuttle, Columbia, launched	•Fukui and Hoffman Nobel Prize for quantum chemistry	•IBM Personal Computer, PC, with MS-DOS
1982	•Falklands War: Britain defeats Argentina	•Agreement to curb ozone-damaging CFCs	•First artifical heart transplant	•Mike Freedman's math for four-dimensional space	•CDs appear •First PC computer "clone"
1983	•U.S. and Caribbean troops invade Grenada	•Global warming, acid rain are world news	•CDs first go on sale •AIDS virus isolated	•Walther Ghering discovers "homeobox" gene in worms	•U.S.: Satellite TV direct to homes in Indianapolis
1984	•New Zealand declared a nuclear-free zone	•AIDS virus, HIV, identified	•Apple Macintosh launched	•Alec Jeffreys develops genetic fingerprinting	•Camcorder for "home movies"
1985	•USSR: Gorbachev named Communist Party secretary	•U.S.: "Star Wars" space defense plan	•U.S.: Tele-shopping	•Clive Sinclair's C5 battery tricycle-car — fails	•Desktop publishing, Apple and other computers
1986	•Chernobyl nuclear disaster	•Challenger shuttle disaster	•Voyager 2 finds 10 more moons of Uranus	•Rutan and Yeager's non-stop around-the-world flight	•Lasers treat clogged arteries in the heart
1987	•Black Monday stock market crash	•Genetic engineering: fast-growing "superfish"	•First flight of fly-by-wire A320 plane	•Madrazo's new treatment for Parkinson's disease	•DAT, digital audio tape •4-wheel drive
1988	•End of Iran-Iraq war •Lockerbie air disaster	•First planets detected outside Solar System	•Color laser photocopier developed	•Stephen Hawking: A Brief History of Time	•Cellular (mobile) phones begin to appear
1989	•China: Tiananmen Square massacre	•"Cold fusion" claims could not be supported	•Voyager 2 reaches Neptune	•Robert Morris jailed in U.S. for computer virus crime	•Game Boy pocket video game

GLOSSARY

atom: the smallest part of a pure substance (chemical element) that can exist naturally. Most atoms are made of three types of even tinier particles called protons, neutrons, and electrons.

black hole: a place where matter is concentrated into such an unimaginably small space that its density is infinite and it has an incredibly huge gravitational pull.

composite: a structural or engineering material made from various substances or ingredients, including metals, ceramics, and carbon-based fibers, to combine the desired properties of each.

electromagnetic spectrum: a whole range or spectrum of waves consisting of combined electrical and magnetic energy. They include radio and TV waves, microwaves, infra-red, light rays, ultra-violet, X rays and gamma rays.

hydroelectricity: a form of electricity that is generated from flowing water. As the water flows, angled blades that are linked to an electricity generator spin, creating electricity.

nuclear reactor: the main part of a nuclear power unit, where a chain reaction occurs as nuclei (central parts of atoms) of the atomic fuel split and release huge amounts of heat and other forms of energy.

ozone: a form of the chemical element oxygen, but with three oxygen atoms joined to form each molecule, called O_3, rather than two as in the normal oxygen gas molecule, O_2.

wind turbine: a modern version of the windmill, with large angled blades on a tall tower, which converts wind power into electricity. Also called an aerogenerator.

MORE BOOKS TO READ

1970–1980. Our Century (series). Prescott Hill (Gareth Stevens)

1980–1990. Our Century (series). Joanne Suter (Gareth Stevens)

After the Spill: The Exxon Valdez Disaster, Then and Now. Sandra Markle (Walker & Co.)

Black Holes. Heather Couper, Nigel Henbest. (DK Publishing)

Great Discoveries and Inventions that Helped Explore Earth and Space (series). Antonio Casanellas (Gareth Stevens)

The Ozone Hole. Closer Look At (series). Alex Edmonds (Copper Beech Books)

Robots Rising. Carol Sonenklar (Henry Holt & Company)

Solar Power. Energy Forever (series). Ian S. Graham (Raintree/Steck Vaughn)

Teacher in Space: Christa McAuliffe and the Challenger Legacy. Colin Burgess (Bison Books)

The World of Computers and Communications. An Inside Look (series). Ian Graham (Gareth Stevens)

WEB SITES

Anatomy of the Concorde.
cgi.pbs.org/wgbh/nova/supersonic/anatomy.html

Hunt for Alien Worlds.
www.pbs.org/wgbh/nova/worlds/

The Truth about Black Holes.
amazing-space.stsci.edu/blackholes/lesson/index.html

Space Camp for Kids!
www.spacecamp.com/kids/

Due to the dynamic nature of the Internet, some web sites stay current longer than others. To find additional web sites, use a reliable search engine with one or more of the following keywords: *black holes, Chernobyl, cold fusion, energy, fiber optics, global warming,* Mir *space station,* and *space race.*

INDEX

acid rain 5
acrylics 19
AIDS 26
Airbuses 18
aircraft 18, 24
aliens 10
Altair 8800
    computers 22
Alvin 6
Antarctica 17
Apollo spacecraft 12
Apple computers
    5, 22
Ariane rockets 11
artificial hearts 26
atoms 7

Barstow Solar
    Furnace 21
Bhopal, India 16
black holes 7
body imaging 26, 27
Boeing 18
Brown, Lesley 27
Brown, Louise 27

carbon fibers 19
Challenger disaster 17
Chernobyl 16, 17
chlorofluorocarbons 17
Clark, Barney 26
computers 5, 22, 23
Concorde 24
CT scanner 26, 27

deep-sea bed vents 6
de Laplace,
    Pierre Simon 7

Edwards Air
    Force Base 15
Einstein, Albert 7
electric cars 24, 25
electron microscopes 26
energy 20, 21
Exxon Valdez 17

fiber-optics 22

Gagarin, Yuri 12
galaxies 7
Giotto probe 11
gliders 15
global warming 5

Halley's comet 11
Hawking, Stephen 7
HIV 26
Hoover Dam 20
Humber Bridge 19
hydroelectricity 20

IBM PC computers 22
in vitro fertilization 27
integrated circuits 28
Intelsat V satellites 23
International
    Telecommunications
    Satellite
    Organization 23
IRAS 7, 31

Jupiter 11

Kiev, Ukraine 17
Kvant spacecraft 13

landers 10
Landsat satellites 8
LCD (liquid crystal
    display) 29
lithotripsy machines 27
Live Aid 23
Lockheed 18

maglev trains 25
magnets 25
Manarov, Musa 13
Mariner probes 10
Mars 10, 11
medicine 26, 27
Mercury 10
Meteosat satellites 9

microchips 28
Mir space station 5, 13
mobile phones 29
monorails 24
Moon landings 5, 8

Neptune 11
Nimbus satellites 9
Nintendo Game Boy 29
North Pole 9
Northrop F-5
    jet fighters 15
nuclear power 5, 16, 17

oil crisis 20
orbiters 10
ozone hole 5, 17

planets
    see Jupiter, Mars,
    Mercury, Neptune,
    Saturn, Uranus, and
    Venus
pocket calculators 29
pollution 5, 16, 17, 20
Post-it® notes 28
power
    see hydroelectricity,
    nuclear power, solar
    power, wave power,
    and wind power
Progress spacecraft 13

rayon 19
robots 18, 19

Salyut space stations
    12, 13
satellites 7, 8, 9, 23
    see also Intelsat V,
    Landsat, Meteosat,
    and Nimbus *satellites*
Saturn 11
Saturn V rocket 12
Seasat 8
Seveso, Italy 16

Sinclair C5 25
Skylab 12
smallpox 26
solar power 21
Sony Walkman 28
South Pole 9, 10
Soyuz spacecraft 12
space probes 10, 11
    see also Giotto,
    Mariner, *and* Voyager
    probes
space race 12
space shuttle 4, 14,
    15, 17
spacecraft
    see Apollo, Kvant,
    Progress, Soyuz, *and*
    Viking *spacecraft*
space stations 12, 13
    see also Mir and
    Salyut *space stations*
space travel 5, 12, 13,
    14, 15
stars 7
steel 19
Sun 17, 21
Sydney, Australia 19, 24
Sydney Opera House 19

tape recorders 29
Titanic 6
Titov, Vladimir 13
trains 24, 25

Uranus 11

Venus 10
video cameras 5
video games 28
Viking spacecraft 10
Voyager probes 10, 11,

wave power 20
wind power 21, 31

X rays 26, 27